SPOTLIGHT ON GLOBAL ISSUES

CLIMATE CHANGE AND EARTH'S POPULATION

Shannon H. Harts

ROSEN PUBLISHING
NEW YORK

Published in 2022 by The Rosen Publishing Group, Inc.
29 East 21st Street, New York, NY 10010

First Edition

Editor: Theresa Emminizer
Book Design: Michael Flynn

Photo Credits: Cover Glen Allen/The Image Bank/Getty Images Plus; (series globe background) photastic/Shutterstock.com; p. 5 jo Crebbin/Shutterstock.com; p. 6 Gino Santa Maria/Shutterstock.com; p. 7 https://commons.wikimedia.org/wiki/File:Sunset_from_the_ISS.JPG; p. 9 rybarmarekk/Shutterstock.com; p. 10 lavizzara/Shutterstock.com; p. 11 -/AFP/Getty Images; p. 12 Siberian Art/Shutterstock.com; p. 13 Stock Montage/Archive Photos/Getty Images; p. 15 (John Tyndall) UniversalImagesGroup/Getty Images; p. 15 (Guy Stewart Callendar) https://en.wikipedia.org/wiki/Guy_Stewart_Callendar#/media/File:GSCallendar1934.jpg; p. 17 https://commons.wikimedia.org/wiki/File:Charles_David_Keeling_%26_GW_Bush_2001.jpg; p. 18 Dennis Cook/AP Images; pp. 19, 24 NurPhoto/Getty Images; p. 20 nicostock/Shutterstock.com; p. 21 Spencer Platt/Getty Images; p. 23 Sean Gallup/Getty Images; p. 25 Mizzick/Shutterstock.com; p. 26 John D Sirlin/Shutterstock.com; p. 27 David McNew/Getty Images; p. 29 kwest/Shutterstock.com.

Library of Congress Cataloging-in-Publication Data

Names: Harts, Shannon H.
Title: Climate change and Earth's population / Shannon H. Harts..
Description: New York : Rosen Publishing, 2022. | Series: Spotlight on global issues | Includes glossary and index.
Identifiers: ISBN 9781725323094 (pbk.) | ISBN 9781725323124 (library bound) | ISBN 9781725323100 (6 pack)
Subjects: LCSH: Climatic changes--Juvenile literature. | Climatic changes--Effect of human beings on--Juvenile literature. | Global warming--Juvenile literature. | Nature--Effect of human beings on--Juvenile literature.
Classification: LCC QC903.15 2022 | DDC 363.738'74--dc23

Manufactured in the United States of America

CPSIA Compliance Information: Batch #CSR22. For further information contact Rosen Publishing, New York, New York at 1-800-237-9932.

CONTENTS

CHAPTER ONE

RISING POPULATIONS, DISAPPEARING SPECIES

Earth's human population is approaching 8 billion people. From the planet's highest peaks to its greatest depths, humanity's impact is becoming more evident than ever. This is partly due to climate change, which is putting the very systems humans depend on for survival at risk.

As the human population continues to grow, animal species are disappearing. The Florida panther, which can grow to around 7 feet (2.1 m) long, is an efficient predator that's perfected the art of gracefully sneaking up on its prey. Despite its many remarkable adaptations, this top predator is on the brink of extinction. That makes it one of the 30,000 or so species a year that humans are driving to extinction. In fact, scientists say the world is in the midst of its sixth mass extinction, and humans are the root cause.

Florida's human population has nearly doubled in the last 30 years, but the Florida panther's population has dropped. In 2017, the Florida Fish and Wildlife Service estimated there were only around 120 to 230 panthers left in Florida, making it one of the world's rarest mammals.

The rising human population and declining panther population is a perfect example of how people are affecting the natural environment. Luckily, many of the planet's top engineers and scientists are making strides to address climate change and people's influences on the planet. Activists are also using their powerful voices to work against the devastating consequences of climate change even as those changes unfold.

Climate change is causing Earth's ice caps to melt, which causes sea levels to rise. This flooding could destroy the Florida panther's low-lying habitat in southwestern Florida.

CHAPTER TWO

EARTH'S CHANGING ATMOSPHERE

Earth is unique among the planets in our solar system. Earth's temperature and mix of chemicals provide the perfect ingredients for life. These unique ingredients haven't been found anywhere else in the expansive Milky Way galaxy.

Earth has been around for about 4.5 billion years. Within that time, it's undergone many changes. During ice ages, vast ice sheets covered huge areas of land. Species have come and gone. Life affects the planet's climate. In turn, the climate shapes how species are able to survive.

Humans have impacted the climate by changing the delicate balance of gases in our atmosphere. The atmosphere is about 78 percent nitrogen, 21 percent oxygen, and less than 1 percent other gases such as carbon dioxide (CO_2).

Fossil fuels are fuels that form in the earth from dead plants or animals. Humans burn fossil fuels, such as coal and natural gas, for energy and during industrial activities. This releases carbon dioxide into the atmosphere. Within the last 150 years, the amount of CO_2 in the atmosphere has increased from 280 parts per million to 400 parts per million. Other greenhouse gases, such as methane, water vapor, and nitrous oxide, are also rapidly increasing due to human activities.

In 2017, CO_2 made up about 82 percent of U.S. emissions. Methane was the second most common gas emission, accounting for about 10 percent. Methane is mainly produced from the decay of organic matter in landfills and from agricultural activities, especially raising livestock.

This photo shows the layers of Earth's atmosphere.

The correct balance of atmospheric gases is essential to life on Earth. Scientists fear that the increasing levels of CO_2 and other greenhouse gases could have a domino effect on Earth's climate. This means that the consequences of greenhouse gas emissions could lead to further climate problems.

A climate tipping point is a threshold that, when passed, can lead to large-scale changes to the climate system. For example, the melting layers of frozen ground in Earth's polar regions could unleash large stores of greenhouse gases from land called permafrost. This would intensify, or worsen, the problems caused by greenhouse gas emissions and climate change. The rapid growth of the human population is believed to be fueling these climate tipping points.

Since AD 1, the human population has grown from 200 million to nearly 8 billion. However, populations aren't growing at the same rate everywhere. Certain populations have much higher greenhouse gas emissions than others.

Populations are growing fastest in the Middle East and Africa. Asia, which is home to around 60 percent of the world's population, is the highest emitter of greenhouse gases. The United States is the second-highest emitter of greenhouse gases, producing 5.3 billion tons (4.8 mt) of CO_2 a year.

Although less-developed countries emit lower levels of CO_2, the consequences of climate change may hit them the hardest. This is just one example of climate justice issues that are becoming increasingly common as Earth's climate changes.

In Alaska, methane gas bubbles in lakes are becoming more common as climate change causes frozen permafrost to thaw.

The human population is now large enough to impact nearly all the cycles that make Earth's land and climate habitable.

The water cycle is the most important system that climate change impacts. The water cycle involves water evaporating from lakes, oceans, and other bodies of liquid water and turning into clouds through condensation. Water then returns to Earth's surface in the form of precipitation, such as rain and snow.

Climate change is causing Earth to heat up. This is a problem because evaporation happens more when it's warmer. Earth's warming climate adds more water to the atmosphere, causing heavier precipitation. Heavier precipitation may lead to downpours and flooding.

These changes in the water cycle have contributed to several recent disasters. In 2017, Hurricane Harvey hit southeastern Texas and claimed 88 lives. In 2018, Hurricane Florence hit North Carolina and claimed around 42 lives.

HURRICANE HARVEY, 2017

Climate change is also causing more intense droughts, or periods of time when there is little to no rain. The most extreme drought in California's history was from 2012 to 2017. Climate change is also causing too much plant transpiration. Transpiration is the process by which plants give off water vapor through openings in their leaves. Increased transpiration causes plants and soil to lose water and dry out.

These changes to the water cycle are causing something called "precipitation whiplash" or "drought and deluge." As these names suggest, this effect involves long dry spells followed by periods of brief and heavy precipitation that leads to flooding.

Heavy rainfall can lead to flash floods. These are becoming increasingly common in many parts of the world, such as India, which is shown here in August 2018.

CHAPTER THREE

EXAMINING THE GREENHOUSE EFFECT

Although climate change is often seen as a modern issue, experiments as far back as the early 1800s helped show that greenhouse gases could contribute to a warmer world. Scientists conducted these experiments to answer a question: could gases generated by humans, such as carbon dioxide, insulate Earth?

In 1824, Joseph Fourier described what we know today as the greenhouse effect. The French physicist and mathematician wanted to know how the temperature of our planet is regulated. Surfaces heated by the sun emit thermal energy. It seemed to him that Earth should be an icy planet because this returns the sun's energy to space.

Fourier concluded that Earth's atmosphere plays a role in regulating its temperature. In fact, without its atmosphere, Earth would be around 60°Fahrenheit (33°Celsius) cooler overall.

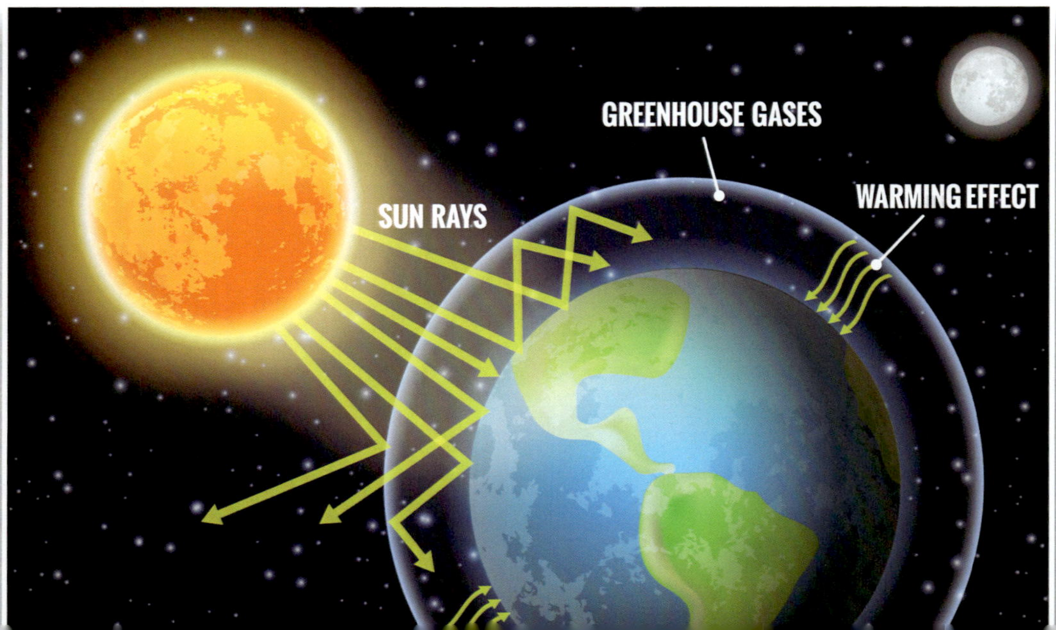

Although Joseph Fourier died on May 16, 1830, his work on the greenhouse effect proved to be key to the scientific understanding of climate change.

Fourier found a simple way to describe how the atmosphere works to keep Earth warm. He used a box with a glass cover. In this well-known experiment, as sunlight entered the box's glass covering, it heated the box. Sunlight passes easily through Earth's atmosphere like sunlight through the box's glass covering. It then warms Earth because it can't easily escape the atmosphere again.

Later experts questioned this experiment and said it was a bit too simple, mainly because radiation heading back to space isn't trapped by Earth's atmosphere, it's absorbed. Still, Fourier was on the right track.

In the 1860s, Irish scientist John Tyndall studied the role of greenhouse gases in man-made climate change. His experiments found that coal gas, which has CO_2, methane, and hydrocarbons, is very good at absorbing heat. Tyndall found CO_2 in particular is able to absorb the sun's radiation like a sponge.

Swedish chemist Svante Arrhenius was next to offer ideas about Earth's climate. In 1895, he took a different approach. He wondered if less CO_2 in the atmosphere might lead to a cooler planet. Arrhenius specifically looked at volcanic eruptions, which produce a large amount of CO_2. His experiments showed that by cutting CO_2 emissions in half, Earth could cool by as much as 9°F (5°C).

While these experiments were going on, greenhouse gases continued to pour into the atmosphere as a result of the Industrial Revolution, which began in Great Britain during the later half of the 18th century.

Use of steam-powered machines increased, changing economies around the world. However, since these revolutionary steam engines often were powered by coal and emitted high amounts of greenhouse gases, the Industrial Revolution is often seen as the beginning of major climate change.

In the 1930s, British engineer Guy Stewart Callendar observed that the northern Atlantic region and the United States seemed to have warmed significantly following the Industrial Revolution. Callendar believed that the warming of the planet was already well underway. His calculations motivated governments to keep an eye on CO_2 levels.

Callendar's ideas about climate change were met with doubt by some other scientists of the time.

JOHN TYNDALL

GUY STEWART CALLENDAR

CHAPTER FOUR

THE KEELING CURVE

One of the most significant developments in climate change research came from data collected in Hawaii starting in 1958. That year, American climate scientist Dr. Charles Keeling found a way to record CO_2 levels with data collected at the Mauna Loa Observatory.

Keeling used the information collected at this observatory to create a graph that shows what's now called the "Keeling Curve." The graph shows seasonal and annual CO_2 levels in the atmosphere starting in 1958. The graph looks like a steep, saw-toothed upward slope. It shows the increase in CO_2 levels every year with slight seasonal dips. These dips show the "greening" of the Northern Hemisphere each spring when plants take in more CO_2 through photosynthesis.

This graph is the longest uninterrupted measure of carbon dioxide levels in the middle layers of the troposphere, the lowest level of the atmosphere. Scientists can compare the records with levels of carbon dioxide trapped in ice cores. Since 1900, during the second phase of the Industrial Revolution, levels of carbon dioxide have steadily increased. During this time period, there was significant industrialization, including increasing automatic operations. These often rely more heavily on fossil fuels.

In this photo, Dr. Charles Keeling receives the Medal of Science from President George W. Bush in 2002.

Other numbers show another steady increase along with the rise in CO_2 levels in the atmosphere—the rise in the human population. In Great Britain, where the Industrial Revolution originated, it's estimated the population doubled from 1750 to 1850. Improved medical technologies and decreasing death rates caused similar population swells in many other parts of the world.

CHAPTER FIVE

A GLOBAL CONCERN

People really started to feel the increase in greenhouse gases in the atmosphere in 1988. That summer was the hottest on record at the time. The United States in particular had many devastating wildfires and droughts related to the high temperatures.

That year marked a major turning point in how the world viewed climate change. In 1988, NASA scientist Dr. James Hansen expressed his concern about climate change to the U.S. Congress. Hansen brought dice with him when he testified. He used them to show his calculations about the probability of climate change causing hotter summers.

Dr. James Hansen, director of NASA's Goddard Institute for Space Studies in New York, testifies before the Senate transportation subcommittee on Capitol Hill in 1989.

One of the dice showed a 2/6 chance of warmer-than-average summers from 1950 to 1980. Two sides of the die were colored red to represent this. Hansen said the other die represented summers in the 1990s. Each of the sides were colored red to represent the "loaded dice" scenario of more warmer-than-average summers as CO_2 emissions continue to rise. These props went with powerful testimony. Both had a stunning impact on political and public opinion regarding climate change.

In 2000, Al Gore, a Tennessee native who served as vice president from 1993 to 2001, ran for president. Addressing climate change was central to his candidacy. Gore won the nation's popular vote, but he lost the electoral vote. Since then, he has continued to advocate for efforts to address climate change. In 2007, he was part of a group that won a Nobel Prize for climate change work.

When Dr. James Hansen testified before Congress in 1988, he said he was "99 percent sure" that the world was in the midst of human-created climate change.

The following year, the United Nations—the international organization that aims to promote peace and progress worldwide—got to work. It formed the Intergovernmental Panel on Climate Change (IPCC). The mission of the IPCC is to provide a scientific view of climate change. The organization also gathers information on climate change's effects and risks and the ways people can adapt to these changes.

While the IPCC doesn't conduct its own research, there are 195 member countries—including the United States—and thousands of people worldwide working to further its mission. The organization's primary goal is to provide expert scientific information on areas of climate change that need to be addressed and to help guide government policies.

On April 22, 2016, U.S. Secretary of State John Kerry signed the Paris Climate Agreement—while holding his granddaughter.

Another major step in addressing climate change came in 2015. In that year, many countries came together to create the Paris Climate Agreement. This agreement centered on the member countries working to prevent the global temperature from rising more than 3.6°Fahrenheit (2°Celsius) from the mark set before the Industrial Revolution. The United States, under President Barack Obama, was one of these countries.

While that temperature change may seem insignificant, it carries a remarkable range of consequences. These include more frequent and deadly heat waves, more intense storms, more severe droughts, and rises in sea level that put cities and communities worldwide at risk. In fact, the Paris Climate Agreement specifically aims to keep rising temperatures "well below" this level.

CHAPTER SIX

POLITICAL CHALLENGES

In October 2018, the IPCC released a report saying that, to avoid the worst consequences of climate change, the world should limit global warming to an even lower target: 2.7°F (1.5°C).

The report, which referred to more than 6,000 scientific works, warned that a warming of 3.6°F (2°C) posed much higher risks. These include deadly weather events, the near-disappearance of all of the world's coral, and summers where the Arctic is ice-free, causing a dangerous rise in sea levels. In addition, millions more people would be pushed into poverty. The report also stressed that the world has only about 12 years to curb greenhouse gas emissions enough to avoid these effects.

Some political leaders, however, didn't heed the warnings of this report. In 2017, U.S. president Donald Trump announced that the United States would withdraw from the Paris Agreement. He stated that it put the needs of other countries ahead of those of the United States. However, his comments were met with protest. In fact, about 600 businesses, including Gap, Nike, and Johnson & Johnson, signed a declaration in support of the Paris Climate Agreement.

Choosing to withdraw from the Paris Climate agreement could have serious consequences. As the report suggested, there is a limit on how much time the world has to curb the effects of greenhouse gases. A future president could rejoin it at a later time. However, they would have to commit to picking up where the United States left off.

President Trump's decision to take the United States out of the Paris Agreement sent shock waves around the world. This photo shows people gathered in protest outside the U.S. Embassy in Germany on June 2, 2017.

CHAPTER SEVEN

DANGERS OF CLIMATE CHANGE

The many numbers swirling around the issue of climate change can be confusing—but there are many physical changes that people can already see. The Arctic is one of the places that's been impacted most. There, the amount of sea ice changes with the season. From space, it's clear that ice in the Arctic has shrunk significantly since 1979.

In Iceland, the impact is so significant that a glacier that had covered about 6.2 square miles (16.1 square km) in 1890 had shrunk so much by August 2019 that it was declared "dead." People even held a funeral to draw awareness to disappearing glaciers.

This image from São Paulo, Brazil, in February 2019 shows the deadly power of a storm surge, which can wash away cars.

Melting glaciers mean sea levels are an increasing threat for the nearly 40 percent of U.S. people living on coasts. Globally, 8 out of 10 of the world's largest cities are coastal. Satellites have been tracking the rise in global sea levels since 1993. In 2018, the average level was 3.2 inches (8.1 cm) above the 1993 average.

This additional "background" water can lead to significantly more damage from storm surges that push farther into cities during major hurricanes such as Katrina and Sandy.

All the fresh water entering oceans from melting glaciers may also be slowing the ocean "conveyor belt." This belt is a system of currents that affects weather around the planet. Slowing it down could cause the Northern Hemisphere to become colder, Indian and Asian monsoon areas to become drier, and North Atlantic storms to be more intense.

Climate change is also fueling another major concern: deadly wildfires. Numerous studies show that wildfires are becoming more common and severe in many parts of the world. Warmer temperatures caused by climate change dry out soil and cause snow to melt sooner, making plants far more likely to catch on fire.

In addition, shifting meteorological patterns mean there's less precipitation in many areas. The wildfire season in the United States is about three-and-a-half months longer than it was just decades ago. In 2018, a wildfire destroyed 95 percent of the California town of Paradise and killed 85 people.

Climate change is also causing the expansion of deserts, a process known as desertification. Dust storms have increased around 25 percent due to desertification. They can cause many health concerns, including breathing issues such as asthma and pneumonia.

A fire devil or fire whirl forms as firefighters battle a California blaze in 2017.

These issues have already made thousands of people in sub-Saharan Africa, South Asia, and Latin America climate refugees. That number is expected to rise by about 140 million before 2050.

Climate change also causes droughts that are stressing urban water supplies. In 2018, Cape Town, South Africa, which has about 4 million people, nearly ran out of water after a drought. Citizens conserved water by taking shorter showers. The government created a citywide map of water-use levels so residents could compare their water usage to their neighbors.

CHAPTER EIGHT

WHAT'S BEING DONE?

The world's population is projected to reach 11.2 billion people by 2100. People are looking at answers including technology and public education to address the problems of rising populations and climate change.

Family planning means educating families to help them plan how many children they'll have and to make sure they have access to birth control. This is an effective way to lower the carbon emissions of growing populations.

As far as technology, carbon capture and storage has shown promising results. This technique "catches" greenhouse gases when they're emitted and stores them. In Iceland, scientists made a major breakthrough in May 2019, when carbon dioxide emissions from a geothermal energy plant were stored underground for the first time in history.

Energy from renewable sources such as the wind, the sun, and the tides provides ways for humans to move away from fossil fuels. In 2017, solar energy made up around 10 percent of total energy generated in the United States for the first time. In Ireland, wave energy generated around 0.5 megawatts of electricity in September 2019—that's enough to power around 500 homes. Thanks to government incentives, Ireland is spending more on renewable energy technology.

However, many governments are slow to invest in these new technologies because of high costs. It's estimated the United States spends around $37.5 billion a year on the fossil fuel industry.

HOW YOU CAN HELP

INDIVIDUAL ACTIONS	IMPACT
Shortening showers by 5 minutes for one month	Save around 3 trillion gallons (11.3 trillion L) of water and water bill costs
Reduce meat consumption	Save around 25 percent of individual carbon emissions
Line-dry clothes instead of using the dryer once a week	Save the equivalent energy of turning on 225 light bulbs for an hour
Vote for government leaders who have plans to curb and adapt to climate change	Have a part in shaping the larger systems that determine your future and that of future generations

CHAPTER NINE

TAKING A STAND

Despite dire warnings from organizations such as IPCC, carbon dioxide emissions hit a new high in 2017 and 2018. That followed a three-year timespan in which they remained mostly flat.

In August 2018, Greta Thunberg, only 15 at the time, decided to do something about it. One day, instead of going to school, Thunberg decided to stand with a homemade black-and-white sign in front of the Swedish parliament building. She started the Fridays for Future movement, which encourages students to go on strike every Friday to demand governments make changes that address the climate crisis. Thunberg's movement now involves around 100,000 students worldwide. She has shown that the actions of one person can make a difference.

Xiuhtezcatl Martinez is another young environmental activist who is using his voice to demand that world leaders make changes. Martinez began speaking out about climate change when he was just 6 years old. He's now the director of a group called Earth Guardians. Martinez and other like-minded young people are part of a lawsuit against the federal government. They argue that failing to take action to address climate change is denying future generations the constitutional right to life.

While the issues of climate change and overpopulation can seem overwhelming, it's important to remember each step toward a more sustainable world makes a difference.

GLOSSARY

activist (AK-tih-vist) Someone who acts strongly in support of or against an issue.

advocate (AD-vuh-kuht) A person who argues for or supports a cause or policy.

climate justice (KLY-muht JUH-stuhs) The intersection of environmental issues and the racial, social, and economic issues they cause.

climate refugee (KLY-muht REH-fyoo-jee) Someone who has to leave an area because of climate-linked issues.

devastating (DEH-vuh-stay-ting) Causing great harm or damage.

emitter (eh-MITT-ur) Something that releases something.

evaporate (ih-VAA-puh-rayt) To change from a liquid into a gas, or vapor.

galaxy (GAA-luhk-see) Any one of the huge groups of stars that make up the universe.

greenhouse gas (GREEN-howz GAS) A gas in the atmosphere that traps energy from the sun.

habitat (HAA-buh-tat) The natural home for plants, animals, and other living things.

hydrocarbon (HY-droh-kahr-buhn) A substance that contains only carbon and hydrogen.

incentive (ihn-SEHN-tihv) A reason to do something.

insulate (IHN-suh-layt) To add material to something to stop heat, electricity, or sound from going into or out of it.

meteorological (mee-tee-uh-ruh-LAH-jih-kuhl) Having to do with the science that involves weather and Earth's atmosphere.

monsoon (MAHN-soon) The summer rainy season in Southeast Asia, or the rain that falls during this season.

nitrous oxide (NY-truhs AHK-syd) A colorless gas that is a pollutant and greenhouse gas.

permafrost (PUHR-muh-frost) A soil layer that is always frozen in very cold regions of the world, such as the Arctic.

INDEX

PRIMARY SOURCE LIST

Page 15
Guy Stewart Callendar. Photograph. Taken 1934. Now kept at the University of East Anglia Archive.

Page 17
Charles Keeling receives the Medal of Science. Photograph. Taken June 13, 2002. Now kept at the National Science Foundation.

Page 21
U.S. Secretary of State John Kerry holds his granddaughter while he signs the Paris Climate Agreement. Photograph. Spencer Platt. Taken April 26, 2016. New York, New York.

WEBSITES

Due to the changing nature of Internet links, Rosen Publishing has developed an online list of websites related to the subject of this book. This site is updated regularly. Please use this link to access the list: www.powerkidslinks.com/SOGI/climatechange